PAST WARS

and

OLD WARRIORS

Jim - See pg. 28
where you are mentioned
Hope you enjoy the
rest of the book too.

by

George R. Conger

PAST WARS and OLD WARRIORS

The author can be contacted via e-mail:
DrGRConger@comcast.net

ISBN (978-0-615-36224-3)

Printed in USA by 48HrBooks
(www.48HrBooks.com)

Dedication

This book is dedicated to past, present and future Navy Plane Captains and Parachute Riggers because they look out for the likes of us.

PREFACE

I started out intending to write a light-hearted little book of recollections. The emphasis was to be on military matters and the people involved in those kinds of things. After some months I came to the chapter dealing with the obfuscation that surrounds service personnel lost in combat. That was some time after the Virtual Vietnam Memorial Wall arrived in my collection of e-mails; sent from a friend whose name I can't recall. It was one of those interesting items you know you ought to keep. This one I was sure I'd like to browse someday, so I placed it in my computer under "FAVORITES."

As I was cogitating on where to begin that "obfuscation" chapter. I was thinking that for more than thirty-five years I knew very little about Harley Hall and the mystery surrounding his disappearance. I knew I wanted to say something about him but as I tried to write something relevant I realized that I didn't have anything new to say. That's when I remembered my list of "FAVORITES." I clicked on The Wall. And to my pleasant surprise I discovered several words of remembrance and tribute included under Harley's name written by those who knew more about Harley than I. Contacting some of those contributors turned my little project around.

Had I been forced to tell about the focus of this book in its first incarnation, I would have said: ‘Simplicity.” And in it I was going to introduce the principle of Occam’s razor . . .

Occam's razor (also spelled **Ockham's razor**) is a principle attributed to the 14th-century English logician and Franciscan friar William of Ockham. The principle states that the explanation of any phenomenon should make as few assumptions as possible, eliminating, or "shaving off", those that make no difference in the observable predictions of the explanatory hypothesis or theory. The principle is often expressed in Latin as the **lex parsimoniae** (law of succinctness or *parsimony*): *entia non sunt multiplicanda praeter necessitatem*, which translates to: *entities should not be multiplied beyond necessity*. This is often paraphrased as

"All things being equal, the simplest solution tends to be the best one."

(Wikipedia ,the free encyclopedia)

Now that the gist of this book has shifted I retained the "razor" because I still think it is a valuable ingredient in the process of solving problems. Occam's razor helped me arrive at a satisfactory conclusion for myself in the matter of Cdr. Harley H. Hall.

– – – – – – –

<u>NOTE:</u> Later on I will mention a few things about Cdr. Hall but I am convinced his story deserves more attention than I have space for here. So, as this manuscript goes to press, I am beginning another book devoted exclusively to Harley Hall. I hope to present compelling answers with documentation regarding his status since 1973.

Introduction

There is no way to measure the pride that comes with being part of our country's military community. We start out devoted to our branch: Navy, Army, Air Force, Marines, or Coast Guard. Then there comes a time after we have matured when the lines blur between the services. We come to value the contributions made by our comrades-in-arms regardless of the uniforms they wore. We honor their devotion and salute their contributions. Case in point. This past March on a weekend retreat at the National Sports Center for the Disabled in Winter Park, Colorado, organized by the Denver VA Hospital "wheelchair center," I spent an evening talking with Mark, an Air Force medic during the Vietnam war, and Denny, a triple amputee who was an Army Platoon Leader in Vietnam. It felt like we served together. Then I realized why we bonded. We were baked in the same loaf.

WWII in Ft. Lauderdale

I grew up in south Florida during World War II. To avoid a nation-wide panic, our government insisted that the enemy had never set foot on our soil. Really?

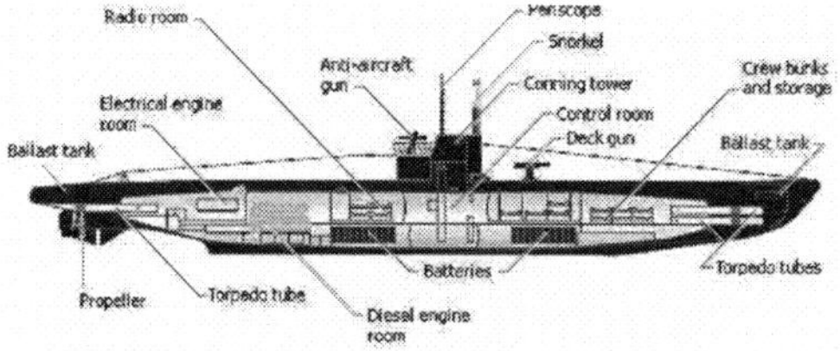

Ft. Lauderdale in 1943 was a small coastal community of about 20,000 year-round residents. Only one road ran east from town to the ocean crossing a turn-style bridge at the Inter-Coastal Waterway, the yachtsmen's thoroughfare from New England to Key West. Just beyond that bridge was a beautiful golden sand beach bordered on the west by a north-south frontage road known as highway Alternate Route 1, or A1A. That beach road paralleled US Highway #1 located a few miles west of the Inter-Coastal Waterway. US-1 was then the major federal thoroughfare that ran south from the top of Maine to the bottom of Florida.

Ft. Lauderdale was a "with it" community. Mindful of the tourist industry, our town council ruled that no private buildings could be erected east of A1A so as to forever guarantee unfettered access to the ocean for everyone. Back then no one could have imagined putting in a barricade to block access to the beach. Putting down barbed wire would have been unthinkable. It took the U.S. Coast Guard to come up with that grand idea. Security required it. And it appeared to work.

Back then there were numerous sightings of German submarines off our coast. Rumors persisted that U-Boats were actually sinking freighters and tankers not far from our beach. One of my friends whose father worked at Port Everglades said that his dad told him about enemy supply ships joining up with Nazi boats and furnishing them with food, ammunition, fuel, fresh clothing and the like. Those stories were made all the more believable when one took into account the fact that Pt Everglades, the deepest salt-water port south of Boston, was situated just a short distance south of the bridge. The Port provided a target-rich environment for axis torpedoes and deck guns. And its oil and gasoline storage tanks must have been inviting as well.

I was an eight-year-old student when these events took place. I think the idea that sparked all of this took shape when I overheard my Dad talking to a neighbor and reminding him that just before the attack on Pearl Harbor a British light cruiser, HMS Orion, intercepted a German freighter, the Arauca, in the waters off Miami Beach. Because we were neutral at the time, the Arauca fled north and put in at Port Everglades to avoid capture. That event caught a lot of local attention. As you might imagine, war-like action close to home made most of the adults around me nervous. Had they know then what we know now, there would have likely been considerably more consternation.

The haunting memory I have of the war was the barbed wire that cluttered my beloved beach. The long roles of wire came above my waist. That stuff seemed to flow endlessly up and down the sand runway that separated the ocean from A1A. The beach was guarded by young Coast Guardsmen dressed in white cotton jumpers and carrying rifles. We presumed those guns were loaded because even us kids knew we were at war. The sentries were spread out all along the beach, perched about five feet off the sand standing on two-by-four boards that constituted lookout towers. They kept watch from A1A to the waterline making sure no enemy sailor swam on shore or crawled under the wire. Naturally, we youngsters made a game of taunting those Coast Guard guys and outrunning them as they chased us off the beach. I wonder now what would have happened if one of those sailors had been successful in the chase. As I recall, not a single buddy of mine was ever caught or shot.

There was a beehive of military activity all around us. The Navy opened a training base south of town at Merle Fogg Airport and designated it the Richmond Naval Air Station. Interestingly, George H.W. Bush received some of his flight training there. You may remember that he was later shot down in the South Pacific and rescued by an American submarine. Mr. Bush later went on to make a rather successful political career for himself and his family.

Port Everglades housed the Coast Guard Station that supplied sentries for my beach. The Navy also had an ammunition storage facility at the Port. Oil tankers steamed in and out regularly filling the storage tanks that fueled our fleet. I have it on good authority that on at least one occasion an American worker did some moonlighting refueling a German sub. Fortunately, he was caught in the act. The U-Boat got away but the traitor was apprehended and, following a quick trial, the scallywag was hung by his neck right there in a metal warehouse building at the Port.

My hometown was fairly important to the defense industry in other ways. We had the HAK Corporation that manufactured artillery shells, Goodwin Tent and Awning that made tents for the Army, and Dooley's Boat Basin and Dry Dock that repaired small boats for the Navy and Coast Guard.

NAS Richmond was primarily assigned patrol duty over Florida waters to search for and destroy German U-Boats. Navy blimps were used extensively for that type of duty and, insofar as I know, the only airship shot down at the time by enemy fire was the K-74. In July 1943 the K-74 was on a routine flight off the Florida Keys looking for enemy sub-marines when it spotted a U-Boat, the

U-134, on the surface of the water in the Florida Straits. The pig boat was charging its batteries. The airship initiated an attack but got more than it bargained for when the sub opened fire with its anti-aircraft guns and downed the blimp.

Whenever we kids were up for some different type of excitement, we would mount our bicycles and ride out to the Naval Air Station to annoy the Navy guards. We were equal opportunity annoyances. We treated Coast Guard and Navy white-hats with equal dis-respect. But bugging the sailors at the airbase was never so much fun as our escapades on the beach with the Coast Guard guys. The Navy had tall fences so we never got inside to do any mischief. But we got to see a lot of airplanes and I think I might have caught a glimpse of George H.W. Bush one after-noon inspecting the propeller of a blue Avenger airplane. If it wasn't him, it was someone who looked an awful lot like him.

I read an account recently written by Brian Murphy entitled Sharks in American Waters. The section I found most interesting was this:

> *It wasn't the US Navy or Coast Guard that controlled America's Atlantic waters in early 1942. It was the U-boats of Nazi Germany. For seven months, from mid-*

January to early August 1942, German U-boats would take control of America's East Coast waters, sinking freighters and oil and gasoline tankers—anything and everything steaming off the coast. Ship by sinking ship, the Nazis achieved a victory over the United States comparable to and even more devastating than the one the Japanese had failed to enjoyed at Pearl Harbor a few weeks earlier. For months, the US Navy came up with a plan to end the slaughter. Meanwhile, the American people were not being told how close they were to disaster.

Murphy reports that by the summer of 1942 the Germans had sunk 233 ships off the east coast of Florida and in the Gulf of Mexico; killing no fewer than five thousand seamen and passengers. He goes on to say that the major event that finally discouraged enemy submarine activity in the area was the work of our American Civil Air Patrol (CAP) which flew mostly small red-and-yellow colored Piper Cubs. A German officer who was taken prisoner was asked why the U-Boats virtually disappeared from our coastal waters by 1945. He replied: "Because of those #%@#& little red and yellow airplanes."

The CAP was formed on December 1, 1941, just six days before the attack on Pearl Harbor. Its mission was to patrol our Atlantic coastline from 21 bases situated from Maine to Mexico. One of those CAP bases operated out of Lantana, Florida, just a few miles north of Ft. Lauderdale. The Germans wanted to disrupt our nation's supply line of fuel, so places like Port Everglades were fertile areas of activity for U-Boats. And where U-Boats congregated, so did those colorful CAP airplanes out of Lantana.

We kids did most of our running around during the summer months because the rest of the time we had to go to school. Mine was Southside Elementary located well west of the Federal Highway US-1 and south of New River, the east-west ribbon of water that bisected Fort Lauderdale. The thing about school I remember most is the baseball field: Southside Park. The ball field was maintained by a man named Fibash. I guess that was his last name. We called him Mr. Fibash. He never answered. He was a strange old duck. But he kept the field in good shape.

One school night my Dad came to my room and announced that we were all getting into the car and driving out to the beach to see something exciting. When we got there we joined several other cars parked along side of the barbed wire. Everyone was looking out to sea. What we saw, way out about three or four miles, was a burning ship. It was big. An oil tanker. While we marveled at the sight, the silhouette of a sub-marine appeared between us and the burning boat. It had to have been the U-Boat that torpedoed the ship. Just like a wolf, the German was circling its wounded prey assessing the damage. I liked to think that it was also rendering assistance to the crew who were abandoning the sinking tanker.

I wondered at the time why no American planes or boats were present. Surely, the military had been notified. After all, my Dad knew about the raid in plenty of time to gather us into the car and to drive out to the beach. Other folks were there, too. Where were our troops?

The next day at school I heard that the U-Boat had been sighted and sunk. I liked hearing that report. Maybe if we had lingered longer we might have seen that action, too.

Anyway, shortly after that memorable evening Mr. Fibash took on a youthful helper. The assistant, Carl, was a good worker and he was much more sociable than Mr. Fibash. But Carl talked like an Englishman. Some of the girls in my class said that Carl was actually a sailor off the German submarine that had been sunk. They thought he had come ashore without being seen by the guards on the beach and, on information supposedly known by all Nazi seamen, he had sought out a person ensconced in our society who would render aid to sailors such as the survivors of that sunken U-Boat. According to the girls, Mr. Fibash was that ensconced citizen right there at Southside Park who was rendering aid and comfort to Carl, whose real name, according to the girls, was Karl.

From that day forward Carl was known as Karl the Kraut. Never to his face, of course.

Karl was the person who started a concession stand at Southside Park. He managed to complete his custodial duties in time to man his food stand during ball games. He sold cold drinks and popcorn that he popped himself in a skillet warmed by an old hotplate. Unfortunately candy was not on his menu,. Chocolate in particular was especially rare during those war years. Come to think of it, rationing touched us all. Rubber tires for cars and bicycles. Gasoline. Cigarettes. Meat. Sugar. And my Mother never did wear nylon hose in those days.

Whenever I got a flat on my bike, I had to repair the inner tube with the meager material provided in a cardboard container that looked a lot like an old pipe tobacco can. I think one of the really wicked things we did as kids back then was to call the local drug store and ask if they had Prince Albert in a can. If they said, “yes.” We’d shout: “Let him out.” How’s that for knee-slapping humor? We all enjoyed a laugh on the south side of the river back in those days.

The river. That river that split the north from the south sides of Lauderdale always intrigued me. I actually believed the Seminole Indian legend that the river appeared mysteriously overnight. That's supposed to be how it got its name: New River. The Indians also said that it had no bottom. A fisherman once let out all 1,000 feet of the line on his deep-sea fishing reel and his sinker never touched bottom. That's what someone told me and I believed it.

Most afternoons on my way home from school I walked along the docks and seawalls along the bank of the river. I enjoyed watching the boats returning from a day of fishing with their assortment of sea creatures. The sharks were always fascinating. And the big grouper. On occasion I would get to help clean some of the fish. My reward was a fish or two for our dinner table. Most of the time, however, I did it just for the fun of it. Between you and me. It was during those times that I developed my love for eating raw fish. An old man of the sea named Zeb introduced me to that delicacy. He's the same guy who taught me how to snorkel.

To this day my friend, Tim Swain, refuses to go fishing with me because I eat too much of the bait.

Zeb cleaned the barnacles off his boat by hand. He would put on a mask with a snorkel. Then he would jump in the river and proceed to rub down the hull of his vessel. It looked like fun, so when he invited me to give it a go, I did. In time I learned to hold my breath for a little over a minute. After going home with wet clothes a time or two I learned to do my diving in my under shorts. I think I was young enough to not cause a stir among the onlookers.

All of the fishermen were friendly to me. Some showed me how to tie basic knots that were essential for boating. One or two let me start up their engines. Getting gasoline never seemed to be a problem for them. I guess that commercial fishermen were seen as essential workers and, like physicians, they got an extra ration of gas.

Zeb was special, however. He fished at night, heading out to sea shortly after sunset and returning to his berth before sunrise. His spot on the river was at the far western end of the seawall, near several old boathouses that had been long ago abandoned and were decaying away like so many other reminders of the boom times of the 1920s when Yankees from Michigan and Ohio tried to make their fortune in this tropical paradise. The Great Depression ruined a lot of dreams back then.

I doubt that Zeb and the other fishermen made much money. They sold their catches to local restaurants and townsfolk and tourists who walked the docks. Since most of the seafarers lived on their boats, I don't imagine that they had much in the way of living expenses.

I made spending money by selling mangos from the back of my red Radio Flyer wagon. The three trees in my yard produced enough fruit to keep me as solvent as a kid needed to be. I got a dime for each mango. My best customers were the guys on the boats at New River and, believe it or not, Karl the Kraut. Karl resold my fruit sliced up in a Coke cup for fifteen cents a serving. He averaged three cups per mango, so Karl did a whole lot better than I. Business is in the genes of those Germans, I guess.

I was young and genuinely oblivious to events going on around me. The war was not a concern and its only impact was more a form of entertainment than fear. But as I went about enjoying the lighthearted things I was doing, sinister events were taking place mere miles from me that would eventually impact my life and the lives of others around me.

Although the Civil Air Patrol had effectively run off most Nazi submarines from the Florida coast, one skilled and tenacious U-Boat remained on station to terrorize and destroy allied shipping. In fact, as late as February 1945 the U-48 engaged a transport ship on route from Cardiff to Port Everglades. The ship was torpedoed within sight of land. This was in all likelihood the final hostile encounter undertaken by the U-48 because later that year she would voluntarily end her illustrious career when she was scuttled off Neustadt, Germany.

This German submarine had several distinctive characteristics but most notable were its twin rudders which made the U-48 uncommonly swift and nimble. The sub was equipped with a deck gun, an anti-aircraft gun, a snorkel that allowed the submerged boat to run at shallow depths while operating its diesel engines, and 21 torpedoes which were launched from four tubes located in its bow and one tube aft. In 1941, when U-Boats left European waters to begin maneuvers in the waters off of

North America, saddle fuel tanks were installed that gave the U-Boats sufficient range to stay on station for a week or longer and then to return to Germany for refueling. These extra tanks contained 33 tons of fuel that permitted the submarine to travel on the surface of the sea for 2,500 nautical miles at a standard speed of 10 knots.

In April 1941, the U-48 sustained serious damage caused during a running battle that resulted in the sinking of a Canadian ship, the S.S. Beaverdale, which had been chartered to the British Admiralty for war service. The Beaverdale, one of the company's state-of-the-art cargo liners, was the first Canadian vessel to fall prey to enemy action. The German submarine limped back to Germany for repairs and refitting. It was probably at this time that the U-48 had an insignia painted on its conning tower: “3 X black cat.” Translation: “this mean black cat has nine lives.” Her record certainly supported that theory.

In calm seas the U-48 normally traveled on the surface at one of four speeds. “Slow” - 2 nautical miles per hour. “Standard” - 10 knots. “Full” speed was 15 knots. And “Flank” speed was 17 knots. Submerged her top speed was 8 knots. The maximum depth of the U-48 was 250 meters (roughly 800 feet).

The U-48 participated in twelve patrols over its four years in the Gulf Stream corridor along the Florida coast. All in all she was credited with sinking 53 allied ships and severely damaged three others. Most amazing is the fact that the U-48 did not suffer a single casualty during all of World War II.

In the warm waters of Florida, to avoid being seen, the sub operated primarily at night. Where she hid during the daylight hours was a constant mystery to the American Coast Guard and Navy.

My favorite theory is that the U-48 had help from someone, maybe a commercial fisherman, who chaperoned the sub up the river at dawn to hide somewhere. Then he guided her back to sea in the evening. The pig boat probably was just barely submerged with its conning tower being the only part of the vessel that could have been seen out of the water. In poor light it might have even looked like a barge the fishing boat was towing. The U-48 was less than 200 feet long, so it could have been hidden up some canal or even far up New River where there was a lot of undergrowth surrounding old piers and docks. I don't know what the draft of a submarine would have been but I know the river was deep enough to contain it. And New River was filled with dark water stained by the tannic acid excreted by the cypress trees lining the shore.

I doubt that I will ever know where that killer boat hid itself. All I know for sure is that at the time when all hostility ceased, not only did the U-48 vanish but

Karl the Kraut disappeared, too. The speed with which he had appeared on the ball field next to Southside Elementary School was matched by the quickness of his departure. I asked Mr. Fibash what he knew about his helper's whereabouts but he acted like he had never heard of Karl.

Walking home that afternoon I passed by Zeb's dock but his fishing vessel wasn't there. Not far from where he normally tied his boat was an old boathouse that looked like it had been recently visited. Out of curiosity I looked inside. The placed smelled of diesel fuel. There were ropes hanging from nails on the wall. An empty gasoline can rested on its side beside the rear entrance. An oil-soaked chambray shirt was lying on the floor. It looked as though whoever had been there had left in a hurry. Then I saw the clue that gave credence to my wild theory. Lying beside the grubby shirt was a paper Coke cup. Inside was a dried mango seed.

Some time later in1945 there was a strange disappearance of five Navy TBM Avengers with 14 aviators aboard who flew out of the Ft. Lauderdale Naval Air Station into the Bermuda Triangle never to return. A PBM Mariner with a 13-man crew was sent out to search for the fliers but It never returned either. I have absolutely no recollection of that bizarre incident. I was ten at time of this crazy event. I guess my attention was directed elsewhere.

By the way. should you be tempted to be critical of what I’ve written or doubt the details I’ve presented; don’t forget, this story was shared based on the recollections of boy who was an eight year old nearly seventy years ago.

KILROY WAS HERE

During the Second World War Kilroy was everywhere. And no one was quite sure who he was or what was meant by his omnipresence. But we were all positively identified with him. We drew his likeness on our book covers at school or on the walls of the locker room.

Kilroy was a strange phenomenon that spread across the country. Everywhere you looked there were pictures of this big nose hanging over what looked like a wall revealing just the top of his head with both hands supporting his snout-like proboscis.

After the war a contest was held to answer our questions about Kilroy. It was run by the American Transit Association through its radio program, "Speak to America." It was a nationwide competition intended to find the real story behind Kilroy. They offered a prize of an actual trolley car to the person who could prove himself to be the creator of Kilroy. Nearly 40 men came forward to make that claim, but only James Kilroy from Halifax, Mass., looked like the real deal.

Jim Kilroy was a 46-year old shipyard worker during the war. He was a checker at the Fore River Shipyard in

Quincy. He'd go around and check on the number of rivets the riveters implanted in the metal frames of the boats being built. Riveters were on piecework and got paid by the rivet. Kilroy would count the rivets and put a little chalk checkmark beside each rivet. That way they wouldn't be counted twice. When Kilroy's workday was over he'd go home and the riveters would go around and erase his marks. Later on, another inspector would come around and count the rivets a second time, resulting in double pay for the riveters.

One day Kilroy's boss called him into his office. The foreman was upset about all the money being paid to the riveters. He told Kilroy to investigate. It didn't take long for James to figure out what was going on.

The spaces he had to crawl through to check the rivets were too tight for him to carry around more permanent kinds of marking material such as a paint can, so he continued to use chalk but beside his chalk mark he added “KILROY WAS HERE” in king-sized letters. He figured that bigger marks would be harder to completely erase. That seemed to work but it wasn’t perfect, so he added the sketch of the character with the big nose. The riveters stopped trying to erase his new, enlarged marks.

Under normal conditions, the rivets and chalk marks would have been covered up with paint. But with a war going on troopships were leaving the Quincy Shipyard so fast that there wasn't time for painting. So, Kilroy's "trademark" was seen by thousands of servicemen who boarded those ships. His message resonated with the soldiers. They picked it up on it and those sketches with the Kilroy message were soon spread all over Europe and the South Pacific. Before the war ended, "Kilroy" had been everywhere.

To the troops Kilroy was a mystery; all they knew for certain was that some clown named Kilroy had "been there first." So, the servicemen began placing the Kilroy graffiti everywhere they landed, proclaiming that Kilroy was already there when they got there.

Before too long Kilroy became the American super-soldier who had "already been there." GIs were creative and always up for a challenge.
Soon the Kilroy message was seen in the most unlikely places. Atop Mt. Everest. On the Statue of Liberty. Scrawled under the Arc De Triumph.

As the war dragged on, the legend of Kilroy grew. Navy Frogmen (underwater demolition teams) were superbly trained swimmers equipped with just mask-snorkel-and-flippers who swam ashore in advance of beach assaults. It was their job to clear the area of obstructions, clearing the way for men and machines to move inland. It turns out that a lot of frogmen made it a point while sneaking ashore on Japanese-held islands to put Kilroy's logo in conspicuous places. The humor and ingenuity of it all caught on. Legend has it that in 1945 an outhouse was built for the exclusive use of Roosevelt, Stalin, and Churchill at the Potsdam conference. Its first occupant was Stalin. When he emerged from the privy it is reported that he asked his aide (in Russian), "Who is Kilroy?"

When he was interviewed for the contest, James Kilroy brought along a few of his co-workers from the shipyard to vouch for him. Needless to say, he won the trolley car and he gave it to his nine children as a Christmas present. They turned it into a playhouse.

Old Ed

My childhood was filled with images of the military. My uncle Andy was a Marine who survived the battle for Guadalcanal. My Sunday school teacher was a bomber pilot in Europe. And, although I doubt that I ever passed this particular old veteran on the beach, if I had it would have been my first sighting of a Medal of Honor recipient. The folks who saw him called him Old Ed. I heard about Old Ed and I was going to include his story in this book until I happened upon this account in a Max Lucado's book: In The Eye of the Storm. Max did a much better job of writing about Old Ed than I.

It happens every Friday evening, almost without fail, when the sun resembles a giant orange and is starting to dip into the blue ocean. Old Ed comes strolling along the beach to his favorite pier. Clutched in his bony hand is a bucket of shrimp. Ed walks out to the end of the pier, where it seems he almost has the world to himself. The glow of the sun is a golden bronze now. Everybody's gone, except for a few joggers on the beach. Standing out on the end of the pier, Ed is alone with his thoughts ... and his bucket of shrimp.

Before long, however, he is no longer alone. Up in the sky a thousand white dots come screeching and squawking, winging their way toward that lanky frame standing there on the end of the pier. Before long, dozens of seagulls have enveloped him, their wings fluttering and flapping wildly. Ed stands there tossing shrimp to the hungry birds. As he does, if you listen closely, you can hear him say with a smile, "Thank you. Thank you."

In a few short minutes the bucket is empty. But Ed doesn't leave. He stands there lost in thought, as though transported to another time and place. Invariably, one of the gulls lands on his sea-bleached, weather-beaten hat - an old military hat he's been wearing for years.

When he finally turns around and begins to walk back toward the beach, a few of the birds hop along the pier with him until he gets to the stairs, and then they, too, fly away. And old Ed quietly makes his way down to the end of the beach and on home.

If you were sitting there on the pier with your fishing line in the water, Ed might seem like "a funny old duck," as my dad used to say. Or, "a guy that's a sandwich shy of a picnic," as my kids might say. To onlookers, he's just another old codger, lost in his own weird world, feeding the seagulls with a bucket full of shrimp.

To the onlooker, rituals can look either very strange or very empty. They can seem altogether unimportant ....maybe even a lot of nonsense. Old folks often do strange things, at least in the eyes of Boomers and Busters. Most of them would probably write Old Ed off, down there in Florida.

That's too bad. They'd do well to know him better. His full name: Eddie Rickenbacker. He was a famous World War I hero. On one of his flying missions across the Pacific, he and his seven-member crew went down. Miraculously, all of the men survived, crawled out of their plane, and climbed into a life raft.

Captain Rickenbacker and his crew floated for days on the rough waters of the Pacific. They fought the sun. They fought sharks. Most of all, they fought hunger. By the eighth day their rations ran out. No food. No water. They were hundreds of miles from land and no one knew where they were. They needed a miracle. That afternoon they had a simple devotional service and prayed for a miracle. They tried to nap. Eddie leaned back and pulled his military cap over his nose. Time dragged. All he could hear was the slap of the waves against the raft. Suddenly, Eddie felt something land on the top of his cap. It was a seagull!

Old Ed would later described how he sat perfectly still, planning his next move. With a flash of his hand and a squawk from the gull, he managed to grab it and wring its neck. He tore the feathers off, and he and his starving crew made a meal of it - a very slight meal for eight men. Then they used the intestines for bait. With it, they caught fish, which gave them food and more bait......and the cycle continued. With that simple survival technique, they were able to endure the rigors of the sea until they were found and rescued. (after 24 days at sea...)

Eddie Rickenbacker lived many years beyond that ordeal, but he never forgot the sacrifice of that first life-saving seagull. And he never stopped saying, "Thank you." That's why almost every Friday night he would walk to the end of the pier with a bucket full of shrimp and a heart full of gratitude.

SOURCE: (Max Lucado, In The Eye of the Storm, pp.221, 225-226)

Eddie Rickenbacker was awarded the Medal of Honor as an Ace in WW I. He later founded Eastern Airlines.

Gored by Gore

In 2009 Irena Sendler died. She was 98 years old. Early in WWII, Irena, a German gentile, got permission from the Nazi authorities to work in the Warsaw Ghetto as a Plumbing-and-Sewer Specialist. She had an ulterior motive for her request. She knew what the Nazi's had in mind for the Polish Jews in the Ghetto and she wanted to rescuer Jewish youngsters before that evil Nazi plan was fully implemented. Irena smuggled infants out of the Ghetto in the bottom of her toolbox. She stored the babies in a burlap sack in the back of her truck. She had a dog she trained to bark when the Nazi guards let her in and out of the Ghetto. Of course, the soldiers wanted nothing to do with the vicious-sounding dog, so they avoided close contact with the dog, the plumber and the truck. The barking dog was an added blessing. It covered whatever noise the youngsters might have been making in the back of the truck.

Before her deeds were discovered by the Nazis, Irena managed to smuggle out over 2,500 Jewish children. After she was arrested, the Nazi thugs beat her severely. Then they broke her arms and legs. She was eventually released.

Irena kept a record of the names of all the children she rescued. She wrote them down on little slips of paper that she placed in a glass jar. She buried it under a tree in her back yard. After the war, she made an effort to locate as many surviving children as she could in the hope that their families might be reunited. Unfortunately, most of the Warsaw Jews had been gassed. Miraculously, all of the children she smuggled out of the Ghetto were either placed into foster homes or they were adopted.

In 2008 Irena Sendler was nominated for the Nobel Peace Prize. She was not selected. Instead, Al Gore won the award. He showed a slide show on Global Warming.

For some twisted reason Mr. Gore was able to convince governments around the world to embrace his unscientific story about humans rather than nature being responsible for climate change. That pro-anthropogenic global warming theory was eventually discarded by most thinking people when e-mails surfaced which were sent between scientists who had fabricated their data and then tried to suppress dissenting views. Dr. Philip Jones of the University of East Anglia was at the heart of those e-mails. Once the smoke cleared Jones admitted that there had never been statistically significant data concerning global warming since 1995. In an interview granted in early 2010 to the BBC, Jones observed: "There was more warming in the medieval period than there has been recently." But he said he is still a pro-anthropogenic believer. Nevertheless, even the United Nations has backed away from its former strong position of blaming man for alterations to our environment. The UN's hype turned out to be an argument based less on scientific truth than on political expediency to further its agenda for promoting sweeping economic rules.

The 2007 United Nation-sponsored report by the Intergovernmental Panel on Climate Change (IPCC) erroneously predicted the Himalayan Glaciers would probably melt away by 2035. The IPCC report concluded: "Climate change is 'unequivocal' and is 'very likely' caused by human activity." In light of what is known today, the IPCC admits that it should have caught the error of its ways with respect to the Himalayan Glacier projection. As a result, public confidence concerning the IPCC has been eroded considerably.

I continue to be haunted by the injustice of Al Gore receiving the Nobel Peace Prize for his hokum and Irena Sendler being passed over in spite of her genuine heroism.

If this were a perfect world, Al Gore would go tomorrow and stand over the grave of Irena Sendler and apologize to her while placing the symbol of the Peace prize on her headstone.

POWs

Have you heard the story of how the British Secret Service, MI-5, during World War II worked with John Waddington, Ltd., the United Kingdom licensee for the American board game: Monopoly, to smuggle maps, metal files, magnetic compasses and various denominations of European currency into Nazi P.O.W. camps? It's true. Those items designed to make escape attempts more feasible were concealed inside "special editions" of the games and sent as "care packages" by the International Red Cross. Those bits of contraband were concealed inside those "special edition" games for use by the captured Brits.

If this is news to you, it's understandable. The whole affair was hush-hush for several years following the war. There were two reasons for maintaining secrecy: the scheme might need to be used again in the event of another war. And there was concern that there might be reprisals against Weddington.

Each game was equipped with several silk maps with "safe houses" clearly marked indicating where shelter could be found. Not only were maps of Germany provided but so were maps of Italy, Sweden, and Norway. Along with the usual supply of play money, real large-denomination Italian, French and German currency was also included.

Using silk for maps was ingenious. Paper maps make lots of noise when opened and they wear out pretty fast, especially when they got wet. Silk, on the other hand, could be folded and unfolded repeatedly without being damaged, and they were quiet. Besides, a silk map could be rolled up into a little ball and tucked into almost any tiny hiding place.

As part of their pre-flight orientation, British and American aviators were briefed on how to identify the Monopoly sets that contained the special items in the event they were captured by the enemy. The clue was a tiny red dot made to look like a normal printing glitch placed on a corner of the “Free Parking” square.

It is estimated that about one-third of the 30+ thousand allied airmen who successfully escaped from German POW camps were helped along their path to freedom by those ingenious Monopoly games.

This wonderfully inspiring story was de-classified in 2007 when still-living workers from Waddington were publicly honored.

This tale adds new meaning to the “Get Out of Jail Free” card, doesn’t it?

OBFUSCATION

From: Obfuscate

Pronunciation: ab-fe skat

Transitive verb:
darken: to make obscure
(e.g., obfuscate the issue)

confuse: (obfuscate the listener)

Obfuscation: noun:
to be evasive, unclear,
or confusing

Source: Merriam-Webster's Dictionary, 2010 edition

– – – – –

For if the trumpet give an uncertain sound, who shall prepare himself to the battle?

- 1 Corinthians 14:8 (King James Version)

Jesse Ventura, former professional wrestler and Governor of Minnesota, came out with a book, *American Conspiracies: Lies, Lies and More Dirty Lies that the Government Tells Us,* in March 2010 detailing a number of government conspiracies. Ventura says he and his co-writer came up with some alternative theories to what people learn from history books and in classroom lectures. One such alternative theory is that people inside the Bush administration knew the 9/11 Tower attacks were coming and they allowed them to happen. This, of course, is so outrageous that it instantly destroyed Ventura's credibility with me.

That's not to say I never doubt the veracity of our government. I do. There are situations that bother me immensely. Missing military personnel is one of the big ones.

The United States Government has a horrible history of abandoning its servicemen previously held prisoner by foreign governments. The Senate Select Committee on POW/MIA Affairs, chaired by Sen. John Kerry, issued a report in 1991 tracing this outrageous practice over a span of some sixty years. It started with an Allied expeditionary force during the Russian Civil War in 1919. That resulted in possibly as many as 130 U.S. soldiers being left behind. They were later written off as "missing, presumed dead."

Then in 1945 there were large numbers of U.S. personnel who had been held in German POW camps who were never repatriated. It seems that when the Red Army overran Germany at the end of WWII, they collected huge numbers of Americans who had been prisoners of the Nazis. They were never returned.

At the time neither the Allied Supreme Command nor the Truman Administration confronted the Soviets concerning those missing Americans. Neither did the former Supreme-Allied-Commander, Dwight David Eisenhower, when he ascended to the presidency of the United States in 1952. To my way of thinking, Ike's attitude toward missing POWs was nothing short of shameful and it soils his memory in my mind. In

1997 a Las Vegas newspaper, the Sun, carried an interview of a former Eisenhower staffer, Philip Corso, who said, "Ike decided the POWs should be given up for dead because the Soviets would never relinquish them anyway."

At the end of the Korean War some 500 known prisoners remained unaccounted for. Speculation has it that they might have been sent to Russia without the Soviets ever admitting that they had them in their custody. Most of our guys were aircrew with considerable technical knowledge of military aircraft. Then, in 1973 after American forces left their combat positions in Vietnam, the U.S. Government demanded that Hanoi return some 5,000 unaccounted U.S. personnel -- a gigantic number compared to the paltry 591 Americans who were actually returned that year. Nothing happened.

Capt. Harley H. Hall

One of those missing men from the Vietnam War was a former shipmate of mine: Cdr. Harley H. Hall. Harley and I were Cadets together in the U.S. Naval School of Pre-Flight at Pensacola in the fall of 1957. I completed my military obligation and returned to civilian life while Harley stayed in the Navy. He devoted himself to flying and rendering distinguished service to his country. He occupies an honored space on the Wall of the Vietnam Memorial. He has the distinction of being the last Naval Aviator lost in that costly conflict.

For years I wondered what happened to Harley. The official Navy word was that he and his RIO were shot down on the final day of our involvement in that war and that the RIO was repatriated later but Cdr. Hall was not.

The Spring 1999 issue of The Hook, the magazine of the Tailhook Association, featured an article entitled, "What Really Happened to Harley Hall." The author was Barrett Tillman. His commentary recounts the official story.

On the last day of official American hostilities in Vietnam, an F-4J from USS *Enterprise* (CVA(N)-65), was shot down near Quang Tri, RVN. It was 27 January 1973. The pilot of *Taproom 113* was CDR Harley H. Hall of Vancouver, Wash., the exec of the VF-143 *Pukin' Dogs* who joined more than 58,000 American servicemen lost in Southeast Asia since 1961. A few others followed before the final debacle ended in 1975, but Harley Hall was the last MIA among Tailhook aviators.

Hall was well known in Naval Aviation. He had been leader of the Blue Angels, flying *Phantoms* in hundreds of performances during his two-year stint from 1970 to the end of the show season in 1971. As XO of the *Pukin' Dogs* (or Dogs as he was wont to say) he was regarded as "a marvel" by no less an authority than retired RADM Tom Brown. CAPT Dan MacIntyre, USN(Ret), himself a Blue, said, "Harley was a good pilot, a good VF driver and a good thinker; a man of honor."

The Last to be Downed - The target that Friday morning was a cluster of North Vietnamese supply vehicles 15 miles northwest of Quang Tri. Working with an Air Force OV-10A FAC, Hall's *Phantom* was struck by AAA fire while pulling off the target. He egressed the area, but the port wing caught fire. There was no option but to eject just offshore. His wingman LT Terry Heath saw two good chutes, watched both of them blown slightly inland and saw Hall run from the area.

Hall's RIO was LCDR Philip A. Kientzler, who had never flown before with the exec. Kientzler landed about 100 feet from his pilot near a village on a small island at the confluence of the Dam Cho Chua and Cua Viet Rivers. About 20 minutes after Taproom 113 went down, the Air Force Bronco, Nail 08, was destroyed by a shoulder-launched SA-7. Though emergency beepers were heard that afternoon and evening, no voice contact was established. Eventually the search was called off.

Phil Kientzler returned from captivity a couple of months later. He told what he could -- he had taken heavy ground fire while in his chute and was wounded in one leg. When he asked the guards about his pilot, he was told the aviator had been killed. Kientzler was fortunate in being captured by North Vietnamese; the two Air Force FACs shot down that morning were caught and murdered by Viet Cong.

POW, Authenticated - No other returning POWs had seen Harley Hall in captivity. However, barely two weeks after his shootdown, his status was changed from MIA to "POW, authenticated." Consequently, the Pentagon listed him as a prisoner for more than six years, and he was retroactively promoted to captain. The apparent contradiction was partially explained when the family learned that North Vietnamese documents indicated he had been captured, though communist representatives in Hanoi insisted they had no information at all about Hall.

In accordance with federal law, Hall was declared deceased on 29 February 1980. In 1995 -- 22 years after the shootdown -- Hanoi returned "partial remains" tentatively identified as CAPT Harley H. Hall, USN. Mary Lou Hall lacked confidence in the government's handling of her husband's case, especially since the chief "pathologist" was merely a mortician. Therefore, she decided to have independent forensic experts examine the "remains": a few tiny bone fragments and three teeth.

That evidence led to an unsettling conclusion - Harley Hall had died in captivity years after the end of hostilities. If that were the case, the logical question was, how many other MIAs and non-returned POWs also had been left behind

Living Proof - Mary Lou Hall and a handful of other families refused to accept the contradictions inherent in U.S. Government evidence about their loved ones missing in Southeast Asia. Still working in San Diego where she raised her daughter and son, Mrs. Hall smiles when she describes herself as "a born troublemaker." Her search for more information led her to sources as diverse as a private laboratory and the Defense Intelligence Agency.

When the teeth and bone fragments were returned, Mrs. Hall demanded an independent examination by a forensic odontologist, who in 1995 confirmed that the teeth were her husband's. The private report stated, "There is no natural phenomenon that tooth No. 5 (and possibly the others) could have been naturally exfoliated with one exception -- CAPT Hall lived far beyond the incident, thereby allowing severe peridontitis to occur and the bone resorption to become so severe that exfoliation could very easily have occurred." Dr. F.N. Powers added that Hall's 1972 dental X-rays showed no periodontal disease. Based on forensic benchmarks, it was estimated that Hall's teeth showed between two and three years of peridontal evidence.

Additional evidence was forthcoming. During a "one time good deal" arranged by the Defense Intelligence Agency, Mrs. Hall and another Vietnam widow gained access to their husband's declassified files. Hall's file was huge -- nearly a foot thick -- and contained much of what Mary Lou had seen before. Intercepted Vietnamese radio messages tracked Hall from battalion to battalion after his shootdown, and there was mention of "a big Blue Angel" being paraded in Hanoi. "Harley was well known because he'd led the Navy's flight demonstration team on an Asian tour in 1970 or 1971," Mrs. Hall explains.

Beyond the previous data, one document shocked Mary Lou Hall: A report that her husband had been interrogated by Soviet officials, presumably in Hanoi. No paper or pens were permitted in the secure reading room, but she memorized the document by date and origin, and then filed a request for its release. Eventually DIA replied that no such report could be found.

The paper trail, combined with the fact that teeth do not constitute "remains," led to the obvious conclusion: There was proof that CDR Harley Hall survived at least two or three years beyond Operation Homecoming.

A question raised at the time concerned the lack of POWs with major injuries. None of the 591 returnees had lost an arm or leg despite high-speed combat ejections, none were mentally impaired and none were obviously traumatized. Then LCDR Jack "Fingers" Ensch, who lost one of his thumbs in his 1972 bailout, was selected for separation from the POW community prior to the mass release, but the senior American officers on site would not permit it, stressing the motto "Unity over self." Consequently, Fingers came home on time.

Mary Lou Hall, a petite blonde with a feisty demeanor, leans forward in her chair. "Yes, my children and I want to know what happened. I don't believe he's still alive today, and frankly, I hope he isn't. The way things turned out, I'd prefer that he was killed on the spot. But what's most important is to know that Harley is remembered and honored with the truth that he's unaccounted for, rather than the unconscionable lies the government has told for so long."

At the close of the 20th century, when at least 13,000 American POWs have been left to their fate by Democrats and Republicans, civilians and military, the question remains: What happened to those men who served the nation that abandoned them?

Beyond that, what can active duty personnel expect of their government after the next major "conflict?" How many prospective Harley Halls are flying military aircraft today? If the lessons of 1919, 1945, 1953 and 1973 are any example, the chilling question answers itself.

Until I began work on this book, Tillman's account was all I knew about Harley's situation. While considering what I was going to say, I accessed an internet WEB page of the virtual Vietnam WALL, the memorial erected in honor of those killed or missing in Vietnam. Attached to Harley's name were several digital comments from those who knew him or were touched by his life. For example:

19 June 2003

It was in the late spring of 1972 at VT-4, NAS Pensacola, while a Naval flight student in basic carrier-quals and ACM in the T-2B, when I had the -- to me - unbelievably great honor of being scheduled to fly some syllabus training flights with the "funner-than-fun" then-CDR Harley Hall! His coach-like charisma was riveting, and he seemed to be continuously holding an impromptu moving-classroom with him wherever he walked or paused - usually with five or six other guys tagging along, competing to ask questions and intently listening to hear his enjoyably mesmerizing and animated answers.

He was the absolute epitome of full-of-life enthusiasm and down-to-earth humor - a

natural-born teacher who spiced our interest in learning all we possibly could about Naval Aviation just by virtue of his personal example of exquisite flying expertise and the uncanny ability to explain it in terms we college boys could enjoy and understand.

CDR Hall was a larger-than-life man's man who had a truly Peter Pan-like bright-eyed empathy for us green-mere-mortals; and yet, he never flaunted his accomplishments. In fact, when I first began flying with him, Harley Hall won my utmost respect, and I just naturally idolized him, without my even realizing at the time he was the former boss of the "Blues" - I just instinctively sensed I was flying with greatness. I was right.

During one of those walking-classroom sessions of his - this particular "period" concerned Air Combat Maneuvering - CDR Hall solidly showed us his firm grasp of human-nature and reality and his own unabashed humility by laughingly interrupting himself during an explanation of a dog- fighting tactic, exclaiming, "Ya' know? We're gonna' teach you guys all these fancy ACM techniques, but as soon as you see your first MIG, all this great knowledge will fly right out the canopy glass, and it'll be 'catch-as-catch-can' from then on!"

I shall never forget him nor that day and his statement as long as I live.

Bless you for your always day-brightening sunshine, Mr. Hall, Sir; Godspeed!

- Max Guiley

- - - - - - - - -

28 Oct 2004

As a "World Famous Pukin' Dog", Harley Hall was the epitome of a Naval Aviator.

I was an AME3 and his Plane Captain on that fateful day in 1973. As I strapped him in, I remember how incredibly proud I felt to be his Plane Captain. My lasting impressions of him were his brilliant smile, impeccable appearance and absolute professionalism.

This was our last sortie that day, and as was rumored, could possibly be the last sortie of the war. Aircraft 102, his bird, was down for maintenance, so we had to go with aircraft 113.

When the shocking news came that he had been shot down, I cried uncontrollably, but I knew in my heart that if anyone could survive, it would be Cdr. Hall.

The "Dogs" were awesome as a Fighter Squadron, and we were that way because of Harley Hall and the men he inspired to be like him.

He made the ultimate sacrifice for his country, and in doing

so, inspired many of us to higher goals.

God Bless Harley Hall, his family, and all of the "World Famous Pukin' Dogs". In my heart, you are still alive.

A shipmate, AFCM(AW)

- Tom Chergi USN(Ret.)

These two laudatory comments were representative. Each gave a glowing account of the man being eulogized. Then I read the following statement from someone who had never met Harley. It blew my hair back.

13 Apr 2007

For over 30 years, the actions of January 27, 1973 at the DMZ have never left my mind. I was there when Harley and Al got shot down and saw them eject and also heard them on Military Air Distress frequency (243.0).

I was approximately 4400 yards away under fire from enemy guns aboard the USS TURNER JOY (DD-951). The first I knew about them being in trouble was a call by CDR Hall over 243.0 saying "Oh s#&*, Mayday, Mayday" and I cut a direction finding bearing on his position in CIC of the TURNER JOY and then ran outside and saw the two of them eject. We watched them dangling in their chutes under fire and immediately started fixing their position. From our viewpoint we thought they landed on an island in the Cua Viet River and later events proved this to be correct. We then started working on getting our five-inch guns into action, laying down cover fire in an attempt to get the now swarming NVA off of them. Listening on 243.0, we heard that CDR Hall had just been captured and he also reported that his RIO had just been killed by the NVA. Apparently Al had just been knocked out and there had been a burst of AK-47 fire in his direction (the last transmission was from Harley, "They just executed my RIO").

We saw the rescue helo trying to get in there to get these guys out (on the last day of the war), but he was under so much heavy ground fire, he apparently could not make it. I can still see that helo trying to get in there. It is still vivid in my mind. I wondered how he could stay in the air with that much ground fire. We did our best to provide more covering fire. Fast Forward! About 23 years later I was home one day in Vancouver, Washington and on the tube they showed CDR Harley Hall, a native of Vancouver, WA and I immediately was flooded with emotions. I was about to retire from the Navy, but all of a sudden these suppressed memories had overwhelmed me and I got in touch with the TV station about Harley and they put me in touch with his sister whom I spoke to for over an hour.

I told her about all the efforts to save the crew, but the fighting was the heaviest of the Vietnam War and it was nearly impossible to retrieve them, but that a major effort had been made. I was glad to get that off my mind. The last day of the war was some of the heaviest combat I had seen and it has never left my mind.

I have visited the Wall in Washington DC and cried at not being able to do more. When the POW's were flown out, the TURNER JOY was one of the covering force ships nearby and I shall never forget that either.

For us who lived this ordeal, all I can say is that a bunch of good Americans did their best to the very end in fighting the Vietnam War and Harley Hall and his RIO were two of them. I am sorry to hear Al has passed away. Back in 1996 I did speak to Al and the helo pilot that tried to get in there. It helped me piece the puzzle together about the events of that day.

Apparently the TURNER JOY fired the last round that impacted at the precise time of the cease fire, which was 0800. At 0759 on January 28th, combat was hot and heavy one minute and about 10 seconds later, it all stopped. The silence was deafening. I shall never forget that. I hope this helps some families out there heal from their loss.

- Jim Chester, Lcdr, USN (Ret)

Commander Chester's account of events that took place on January 27, 1973, casts an entirely different light on the final chapter of Harley Hall's life. And, insofar as I can determine, it is a story that has not yet been recounted publicly. I talked by phone with Jim Chester at his home in Nevada. He is a fascinating man with a distinguished career as a Navy Mustang. He has a distinctive historical role of his own to play relative to the end of hostilities in Vietnam. His riveting explanation of various efforts to rescue Harley held my attention: (1.) an attempted rescue by an American helicopter, (2.) the drop of a SEAL team and (3.) the testimony of a Vietnamese army officer who was on site. Before talking with Cdr. Chester I was unaware that contact had ever been made with on-site eye-witnesses to Harley's shootdown. Obviously, there were enemy soldiers in the area but I doubt that anyone even considered the possibility that one of them might turn up to tell what happened from their prospective. But one did.

Come to think of it, I never saw an after-action report of the incident nor have I seen any comments from Harley's wingman or his RIO who had given his seat to Al Kientzler that day. Of course, for the sake of secrecy the SEAL Team's existence could not be verified let alone the details of their assignment. The rescue chopper surely must have had something relevant to add. But none of that was made public insofar as I know. Now one of the most concrete statements concerning the events on the ground comes from an enemy combatant. Go figure.

In the book to follow, I hope to reproduce Capt. Toan's field journal. I also want to get additional observations from Harley's regular RIO, and his wingman when he went down. I'll seek out the helo driver. And I would like to report on communications with Mary Lou, Harley's widow, as well as Nyla, Al's widow, and other pertinent personnel. **Stand by.**

KIA

Jim Chester managed to read the field journal of the Vietnamese officer, Capt. Nguyen duc Toan, who was on site when the parachutes holding Harley and Al Kientzler deployed. The details reported here are those of Capt.Toan as told to me by Cdr. Chester.

Once he hit the deck, Harley headed for the woods while, some distance away, Al was under attack by enemy ground troops. Capt. Toan was hold up in a bunker on the edge of the field with several soldiers and a nurse. They watched the Americans float to the ground. Raised a religious person with a fanatical respect for life, Toan took several men and headed immediately to the aid of the aviator who appeared to be more vulnerable; the RIO, Al Kientzler. Toan noted that the other American had rushed to the wooded area that appeared to offer him some degree of safe cover. By the time Toan and his men got Kientzler to the bunker and handed him over to the nurse, Toan could no longer see the other American who had run for cover in the woods. Hoping to also rescue him, Toan moved immediately to the woods where he found Harley dead with a wound to his head. Toan and his men buried Harley in a shallow grave close to where he had been shot.

Al Kientzler was wounded and unconscious while Toan was tending to

Harley. Later, when Toan returned to where Kientzler was resting, Al asked him about his pilot. Toan told Kientzler that the other airman was dead.

In April 1988, Nelson DeMille's book, The Charm School, appeared on bookshelves all over America. This is a powerful, well-written but terribly disturbing novel of Cold War intrigue. It takes place in what was then the old Soviet Union,

The plot is straightforward but incredibly disturbing. Two U.S. diplomats working in the Soviet Union stumble upon a horrifying secret. The KGB has created a town as nearly identical as possible to a typical American community peopled by more than 1,000 American MIA pilots who were shot down over Vietnam and spirited away to Russia. The purpose is to set up "Mrs. Ivanova's Charm School" to serve as a KGB academy where the captured pilots are forced to teach aspiring Russian spies how to act, speak, and dress in such a way to pass themselves off as Americans. The KGB will go to whatever lengths required to keep the existence of "The Charm School" a secret.

It's a page-turning thriller of a book. As I recall, it was the first book by DeMille I'd read. I've made it a point to read everything he has written since then. Anyway, I remember at the time wondering if Harley might be held in a similar situation. I, like most other aviators with whom I have spoken, felt that he was alive but stranded somewhere unable to find his way home.

After hearing what Jim Chester had to say, I nearly wept. I realized we were close to having a definitive answer to the mystery surrounding Cdr. Hall's disappearance. I knew that something so important needed verification. One point of proof beyond Toan's journal and Cdr. Chester's view from the Turner Joy could have come from Al Kientzler . It would be nice if Al could talk with Capt. Toan again; preferably at the site of the shootdown. Unfortunately, Cdr. Kientzler died in February 2005 without knowing about Toan 's journal.

According to Chester, Al's widow, Nyla, did meet Toan and he accompanied

Nyla and Al's brother, Gary, to the place where this haunting saga began. Naturally, before his death Al Kientzler had confided in his wife on all the particulars surrounding his F-4J journey that fateful day from its takeoff from the USS Enterprise, to the repeated runs they made on their target, getting hit by enemy fire, ejecting, getting wounded and then being captured. Nyla made the trip to Vietnam with considerable familiarity of the events that took place on that field where her husband landed

According to Chester, Nyla was satisfied with Toan's authenticity. After the tour of the site and seeing the place where Toan buried Harley, Nyla was fairly well convinced that Cdr. Hall was killed the day he and Al landed in that field.

hobnobbing

Every pre-flight class at Pensacola in the 1950s was filled with remarkable young men, I'm sure. However, it's hard to believe that any other class could compare to Class 32-57, my group of 27 cadets. Four retired as Navy Captains. One of them flew with the Blue Angels. Half-a-dozen retired as Captains with major U.S. airlines. And one I know of was awarded the Purple Heart for injuries suffered when his helicopter was shot down in Vietnam. Fortunately, a squad of Rangers was nearby and they rescued **Cdr. Roger Hulson,** our Purple Heart Recipient .

Capt. F. X. Mezzadri

Frank Mezzadri and I met in pre-flight. We kept in touch until his death in March 2002. I had the honor of flying formation with Frank and Harley Hall when all three of us were in the Program driving T-28s out of Whiting Field. Frank and Harley were tight and that was understandable because they were a lot alike.

Frank organized Spud Flight. Once I was invited to join the group and I saw first hand why their futures in Naval aviation were destined to soar far above the rest of us.

"One potato, two potato, three potato, four." That was how we counted off when we rendezvoused over a lake north of Milton, Florida, putting on our clandestine air show.

There had to have been a forth that day but I don't recall who he was. All I remember is that I marveled at those two talented guys destined to make names for themselves in fast movers. In retrospect, I realize I enjoyed a rare honor that day when I was flying loose formation with two future Blue Angels.

Shortly after Frank sewed on his forth stripe he was assigned to the staff of the War College in New Port. He got me included in a gathering of impressive Navy and Marine flag officers discussing the role of the Navy in world affairs. I remember meeting the Secretary of the Navy at the time, John Lehman, who had just made Commander in the Navy Reserve. RADM Lew Chatham, a former Blue, was a guest in Frank's home one evening. We shot the breeze in our stocking feet. Can you picture that? Cornelius A. McGillicuddy's grandson was there, too. RADM Connie Mack, a Naval Reservist from Florida and a nice guy, as I recall.

I was hobnobbing with an impressive group of brass hats thanks to my remarkable friend, Frank. Hanging out with Mezzadri was seldom dull and always fun. I sure miss him.